JN403567

PHOTO GRAPH
BOOK SERIES
OF THE WORLD
INSECTS

―― 세계 장수풍뎅이 컬렉션 ――
The Collection of Rhinoceros Beetles in the world

|주| 커뮤니케이션열림

다윈의 '딱정벌레 모으기' 열정

인류와 자연 진화 과정의 비밀을 밝힌 영국의 생물학자 찰스 로버트 다윈(1809~1882)
다윈은 어린 시절부터 조개껍데기, 새알, 암석과 광물, 곤충 등을 열심히 수집하러 다녔다.
특히 딱정벌레를 놓치지 않으려고 입속에까지 넣었다는 일화는 유명하다.
"살아남는 종은 강한 종도 아니고, 똑똑한 종도 아니다. 변화에 적응하는 종이다."
다윈이 남긴 말이다.

Darwin's passion for 'beetle collecting'

Britain biologist who unveiled the mystery about evolution of mankind and nature
Charles Robert Darwin (1809-1882)
Darwin collected seashells, bird's eggs, rocks and minerals, insects, etc. in his childhood.
An anecdote that he put a beetle in his mouth not to lose is famous.
"It is not the strongest of the species that survives, nor the most intelligent,
but the most responsive to change," said Darwin.

Foreword

by Son, Minwoo

전 세계에 알려진 곤충은 약 100만~110만 종이 알려져 있으며, 해마다 새로운 종이 발견되고 있다. 그 중 장수풍뎅이(Dynastid beetle)가 속해 있는 딱정벌레목 곤충은 가장 큰 분류군을 가지고 있으며, 약 40만종을 차지할 정도로 그 종류가 매우 많고 다양하다. 딱정벌레목의 특징은 두 쌍의 날개 중 초시(初翅, elytron)라 부르는 한 쌍의 앞날개는 딱딱하게 굳어 있어 갑옷을 입은 것과 같이 몸을 보호할 수 있으며, 다른 한 쌍의 뒷날개는 얇은 막 상으로 앞날개보다 크며, 비행시 사용하였다가 정지시에는 앞날개 밑으로 접어 넣어 비행날개를 보호할 수 있는 보다 진보된 기능을 갖고 있다. 바로 이러한 점이 세계각지의 고산이나 평야, 하천과 늪, 지상이나 동굴, 식물체의 내부와 외부, 흙 속 등, 거의 모든 지역에서 적응하고 다양한 종으로 번성할 수 있게 되었다고 본다.

장수풍뎅이의 장수(將帥)는 장군의 의미를 가지고 있다. 영어와 일본어로도 각각 장군을 뜻하는 제네랄비틀(general beetl)과 가부토무시(かぶとむし)로 불리며, 코뿔소를 연상시켜 라이노써로스비틀(rhinoceros beetle)이라고도 한다. 한국에서는 장수의 투구와 머리뿔의 모양이 닮아있어 투구벌레로 불리기도 하였다.

장수풍뎅이는 딱정벌레목(Coleoptera) 풍뎅이과(Scarabaeidae) 장수풍뎅이아과(Dynastinae)에 속하며, 알-애벌레-번데기-어른벌레로 4단계의 성장과정을 거치는 완전변태 곤충으로 수명은 보통 1년으로 현재 8족 201속 1,400종이 알려져 있다. 그들 중 대다수의 많은 종이 동남아시아에 분포하고 있으며, 남아메리카, 아프리카순으로 분포한다.

Currently, approximately 1 million to 1.1 million insect species are known of the world wile many new species are described every year. Coleoptera order, which includes the Dynastid beetle is one of the biggest insect groups of them all with some 0.4 million species falling under it. The order of Coleoptera is characterized by the front pair of wings, called elytron, being hard and functioning as a protective mechanism like a suit of armor while the rear pair, thin like films, is bigger than the other pair. The latter is used only when an insect in this order flies and kept under the front pair of wings for protection when it is not flying, which shows an advanced evolutionary function. This is seen as a reason this order can adapt to and diversify itself successfully in almost every ecosystem worldwide alpine area, plain, stream, wetland, cavern, inside and outside of plants, earth, etc.

The Korean word for the Dynastid beetle contains the meaning of 'a general', while its English and Japanese counterparts 'general beetle and gabutomushi'also have the meaning of 'a general.' The Dynastid beetle is sometimes called the rhinoceros beetle as it resembles a rhinoceros. In South Korea, it is often called 'a helmet bug,' as many Koreans think the beetle's horns look like a horned war helmet.

The Dynastid beetle is classified into the Coleoptera order, Scarbaeoidae family and Dynastinae subfamily. They go through a 4-step complete metamorphic process (egg - larva - pupa - adult) and their lifespan is about 1 year on average. Until now, 8 tribes, 201 genus and 1,400 species of it have been discovered. The majority of them inhabit Southeast Asia, followed by South America and Africa.

Contents

한글명	학명	쪽
섹스푼크타타점박이장수풍뎅이	Cyclocephala sexpunctata	16
버클리장수풍뎅이	Lycomedes buckleyi	18
훔볼드티솥뚜껑장수풍뎅이	Mitracephala humboldti	20
칸토리장수풍뎅이	Dipelicus cantori	22
앞데루스긴뿔장수풍뎅이	Diloborderus abderus	24
핑구이스장수풍뎅이	Blabephorus pinguis	26
아르마투스넓적뿔장수풍뎅이	Ceratoryctoderus armatus	28
빌로바장수풍뎅이	Coelosis biloba	30
판판장수풍뎅이	Enema pan	32
할르투스털장수풍뎅이	Heterogomphus hirtus	34
나씨코르니스장수풍뎅이	Oryctes nasicornis	36
마르타바니장수풍뎅이	Trichogomphus martabani	38
자메이쎈씨스장수풍뎅이	Xyloryctes jamaicensis	40
장수풍뎅이	Allomyrina dichotomus	42
프훼이훼리털장수풍뎅이	Allomyrina pfeifferi	44
아틀라스장수풍뎅이	Chalcosoma atlas	46
아틀라스장수풍뎅이	Chalcosoma atlas	48
아틀라스장수풍뎅이	Chalcosoma atlas	50
키론장수풍뎅이	Chalcosoma chiron	52
몰렌캄피장수풍뎅이	Chalcosoma mollenkampi	54
켄타우르스장수풍뎅이	Augsoma centaurus	56
그란티장수풍뎅이	Dynastes granti	58
티티우스장수풍뎅이	Dynastes tityus	60
헤라클레스리치장수풍뎅이	Dynastes hercules lichyi	62
넵튠장수풍뎅이	Dynastes neptunus	64
벡카리삼각뿔장수풍뎅이	Eupatorus beccarii	66
비르마니쿠스오각뿔장수풍뎅이	Eupatorus birmanicus	68
그라실리코르니스오각뿔장수풍뎅이	Eupatoraus gracilicornis	70
씨아멘시스오각뿔장수풍뎅이	Eupatorus siamensis	72
수키티오각뿔장수풍뎅이	Eupatorus sukkiti	74
크라비게르앞장다리장수풍뎅이	Golofa pizarro	76
피자로앞장다리장수풍뎅이	Golofa pizarro	78
에아쿠스앞장다리장수풍뎅이	Golofa eacus	80
펠라곤앞장다리장수풍뎅이	Megasoma mars	82
악테온도깨비뿔장수풍뎅이	Megasoma actaeon	84
마르스도깨비뿔장수풍뎅이	Torynorrhina flammea chicheryi	86
엘라파스코끼리장수풍뎅이	Megasoma elephas elephas	88
기아스코끼리장수풍뎅이	Megasoma gyas rumbucheri	90
조에르겐세니코끼리장수풍뎅이	Megasoma joergeseni joergenseni	92
테르시테스코끼리장수풍뎅이	Megasoma thersites	94
기데온플로렌시스장수풍뎅이	Xylotrupes gideon florensis	96
프베센스털장수풍뎅이	Xylotrupes pubescens	98
외뿔장수풍뎅이	Pentodon quadridens	100

선진국의 곤충전시장에서는 심심치 않게 곤충도난사건이 벌어지곤 한다. 그만큼 선진국에서는 곤충수집이 활성화 되어 있으며 많은 사람들이 즐기는 취미이기도 하다. 지난 99년 5월 도쿄 토시마의 한 곤충점에서는 일본산 천연 사슴벌레 등 86마리가 없어졌는데 피해액이 물경 약 1억원에 달했다. 한국이라면 도저히 상상할 수 없는 일. 조그만 곤충을 수집하는 사람이나, 훔쳐가는 도둑이나, 또 이를 대대적으로 보도하는 신문 모두 생소한 일일 것이다.

우리는 이러한 생소함에 더해 학습지 성격이 강했던 곤충도감의 성격을 그래픽 모티브로서도 훌륭히 표현되어 색다른 가치를 가질 수 있음을 알리고자 한다. 이런 작은 노력이 얼마나 빛을 보게 될지는 두고봐야 겠지만...

There are occasionally burglaries of insects in an exhibition hall of insects in advanced countries. This indicates that insect collecting is many people's favorite hobby in advanced countries. 86 insects including Japanese stag beetle were stolen at an insect store in Toshima, Tokyo in May, 1999. The amount of damage for the insects was about 100 million won. It will not happen in Korea. Korean people will feel weird about a person who collects small insects, a thief who steals insects, and a newspaper which reports burglary of insects.

We would like to inform that an illustrated insect book is not just learning material, but can have unique value by great expression as graphic motive. I wish that this small effort can come to fruition.

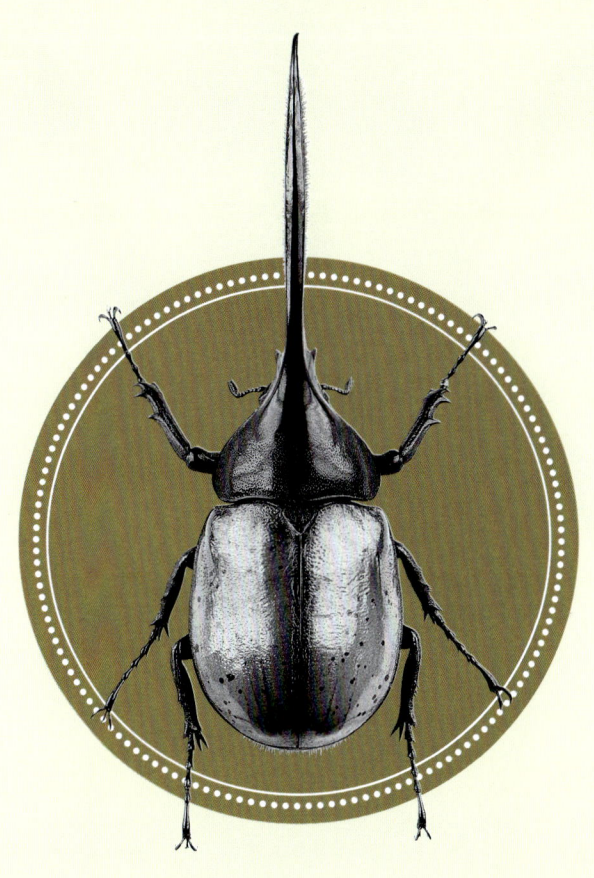

♂
멕시코 산 22mm
(Xicotepec de Juarez, Puebla, Mexico. 2004.8)

Cyclocephala

학 명 Scientific name	**섹스푼크타타점박이장수풍뎅이** *Cyclocephala sexpunctata*
채집국 Collected locality	🇲🇽 멕시코 _ Mexico
크 기 Size	♂ 20-22mm, ♀ 20-22mm

● 채집지 Collected Locality
● 분포지 Distribution

멕시코 Mexico
기아나 Guiana

중앙아메리카의 멕시코부터 남아메리카 북부의 기아나(Guiana)까지 서식한다. 이 종은 *Cyclocephala*속 중 비교적 대형에 속한다. 장수풍뎅이의 큰 특징인 뿔을 가지고 있지 않다. 머리를 제외한 전체의 몸빛이 등황색이며 *C. lunulata*와 비교하여 보다 뚜렷한 검은 반점을 가지고 있다.

Relatively large-sized compared with the others in the genus of *Cyclocephala*, it inhabits Mexico to Guiana. It has no horn, which is one of the most unique characteristics of Rhinoceros Beetles. Its entire body but the head is orange-yellow and its black speckles are more conspicuous than those of *C. lunulata*.

[Distribution] Central America and Guiana

♂
에콰도르 산 40mm
(Enero, Napo, Ecuador. 2002.10.)

Lycomedes

학 명 Scientific name	버클리장수풍뎅이 *Lycomedes buckleyi*
채집국 Collected locality	에콰도르 _ Ecuador
크 기 Size	♂ 30-40mm, ♀ 30mm

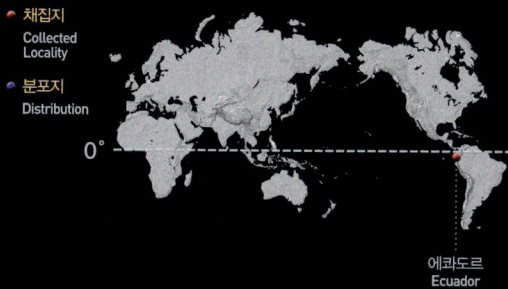

● 채집지 Collected Locality
● 분포지 Distribution

에콰도르 Ecuador

남아메리카 북부 에콰도르의 낮은 산간에 국소적으로 서식한다. 이 종은 머리의 뿔과 가슴뿔이 발달하여 장수풍뎅이와 같은 느낌을 준다. 몸빛은 암회색으로 광택이 없다. 우기에 출현한다.

It lives, dotted, in the low mountain area of Ecuador. This species, as it has developed cephalic and pronotum horns, looks like a type of Rhinoceros Beetle. Its body is dark gray, and does not gloss. It is mainly seen during the rainy seasons.

♂
에콰도르 산 49mm
(Cosanga, Napo, Ecuador. 2004.10.)

Mitracephala

학 명 Scientific name	**훔볼드티솥뚜껑장수풍뎅이** *Mitracephala humboldti*
채집국 Collected locality	에콰도르 _ Ecuador
크 기 Size	♂ 30-49mm

- 채집지 Collected Locality
- 분포지 Distribution

에콰도르 Ecuador · 콜롬비아 Colombia · 페루 Peru · 볼리비아 Bolivia

남아메리카 북서부에 서식한다. 이 종은 몸크기에 비하여 머리의 뿔과 가슴뿔이 작게 발달하여 마치 솥뚜껑과 같은 느낌을 준다. 몸빛은 머리와 가슴은 검정색이고 딱지날개는 암갈색이다. 미약한 광택을 지니고 있다.

It inhabits the northwestern part of South America. As its cephalic and pronotum horns are relatively small and underdeveloped in comparison to its body size, it sometimes looks like the lid of a kettle. Its head and thorax are black while its elytra (hard wings) are dark red. It shows a subtle gloss.

[Distribution] Colombia, Ecuador, Peru, Bolivia

♂
자바 산 36mm
(Java, Indonesia. 1999.)

Dipelicus

학 명 Scientific name	칸토리장수풍뎅이 ***Dipelicus cantori***
채집국 Collected locality	인도네시아 _ Indonesia
크 기 Size	♂ 32-42mm, ♀ 25mm

- 채집지
Collected Locality
- 분포지
Distribution

0°

수마트라 섬 / Sumatra I.　자바 섬 / Java I.　술라웨시 섬 / Sulawesi I.

인도네시아 수마트라부터 술라웨시에 걸쳐 넓게 서식한다. 이 종의 가슴은 화산의 봉우리처럼 움푹 파여져 있고 그 주위로 네 개의 뾰족한 뿔이 발달하여 있고 머리 뿔은 길고 뾰족한 외뿔을 가지고 있다. 몸빛은 검정색이고 강한 광택을 지니고 있다.

It is discovered in Indonesia. Around its thorax, dented like a crater, are 4 sharp horns while there is one, long and sharp horn on its head. Its black body has a strong gloss.

[Distribution] Sumatra Island, From Java Island to Sulawesi Island.

♂
아르헨티나 산 30mm
(Argentina. 2000.)

Diloboderus

학 명 Scientific name	앞데루스긴뿔장수풍뎅이 ***Diloboderus abderus***
채집국 Collected locality	🇦🇷 아르헨티나 _ Argentina
크 기 Size	♂ 20-30mm, ♀ 24mm

● 채집지 Collected Locality
● 분포지 Distribution

파라과이 Paraguay | 아르헨티나 Argentina | 우루과이 Uruguay | 브라질 남부 S. Brazil

남아메리카 남동부에 서식한다. 이 종은 크기에 비하여 가슴이 크고 가슴뿔 또한 두 갈래로 크게 발달하여 있다. 몸빛은 머리와 가슴은 검정색이고 딱지날개는 암갈색이다. 수컷은 광택이 없고 암컷은 미약한 광택을 지니고 있다. 소의 배설물을 먹고 사는 특이한 종이다.

It is mainly discovered in the southeastern part of South America. It has a thorax relatively big in comparison to its average body size, and the large-sized pronotum horn on it branches into two. Its head and thorax are black while its elytra (hard wings) are dark red. Though the male has no gloss, the female shows a subtle one. This species is very unique in that it feeds on cattle's excrement.

[Distribution] Southern Brazil, Paraguay, Uruguay, Argentina

♂
타이 북부 산 32mm
(N. Thailand. 2000.)

Blabephorus

학 명 Scientific name	핑구이스장수풍뎅이 ***Blabephorus pinguis***
채집국 Collected locality	타이 _ Thailand
크 기 Size	♂ 22-32mm, ♀ 25-32mm

동남아시아에 넓게 서식한다. 이 종은 가슴중앙부가 함몰되어있어 양쪽으로 돌기가 발달한 것처럼 보인다. 머리의 뿔은 안쪽으로 굽어져 있다. 몸빛은 적갈색이며 말단부들은 검정색을 띤다. 광택을 지니며 뚜렷하지 않은 점각이 있다.

Living in the SE. Asia, this species has a dent at the center of its thorax, with what looks like protrusions around it. The cephalic horn is bent inward. Its body, glossy and having subtle stripes of stipples, is mainly maroon-colored while its tips are black.

[Distribution] From Srilanka to Thailand, Malaysia, Sumatra Island, Borneo Island, Philippines

♂
술라웨시 산 49mm
(Sulawesi, Indonesia. 2002.)

Ceratoryctoderus

학 명 Scientific name	아르마투스넓적뿔장수풍뎅이 *Ceratoryctoderus armatus*
채집국 Collected locality	🇮🇩 인도네시아 _ Indonesia
크 기 Size	↑ 49mm

● 채집지
Collected Locality

● 분포지
Distribution

술라웨시 섬
Sulawesi I.

인도네시아 술라웨시 섬에 서식한다. 이 종은 가슴중앙부가 함몰되어 있으며 양쪽의 돌기가 전방향으로 발달해 있다. 머리의 뿔은 두텁고 넓으며 말단부는 두 갈래로 갈라져 있다. 몸빛은 검정색의 미약한 광택을 지니며 뚜렷한 세로줄의 점각이 있다.

Inhabiting Sulawesi Island, Indonesia, this species has a subsided thorax center and some protrusions developing into all directions around it. Its cephalic horns are thick and wide with the tips splitting into two. Its black and subtly glossy body has conspicuous stripes of stipples.

[Distribution] Sulawesi Island

♂
콜롬비아 산 39mm
(Narino, Colombia. 2004. 3.)

Coelosis

학 명 Scientific name	빌로바장수풍뎅이 *Coelosis biloba*
채집국 Collected locality	콜롬비아 _ Colombia
크 기 Size	♂ 28-50mm, ♀ 37mm

● 채집지 Collected Locality

● 분포지 Distribution

멕시코 Mexico   콜롬비아 Colombia   아르헨티나 Argentina

중앙아메리카에서 남아에메카까지 넓게 서식한다. 이 종은 가슴 중앙부가 두텁고 넓게 전방향으로 발달하였으며 수컷의 머리뿔은 안쪽으로 굽어 있다. 몸빛은 암적색, 적갈색을 띠며 세로줄의 뚜렷한 점각이 있다.

This species inhabits from Mexico to Argentina. Its thorax center is thick and developed into all directions while the cephalic horn of the male is bent inward. Its dark red and maroon body has some conspicuous stripes of stipples.

[Distribution] From Mexico to Argentina

♂
아르헨티나 산 47mm
(Misiones province, Argentina. 2004. 4)

Enema

학 명 Scientific name	판판장수풍뎅이 ***Enema pan***
채집국 Collected locality	 아르헨티나 _ Argentina
크 기 Size	30-90mm

● 채집지
Collected Locality

● 분포지
Distribution

아르헨티나
Argentina

남아메리카에 넓게 서식한다. 이 종은 가슴뿔이 두 갈래로 나뉘어져 전방향으로 굽어져 있으며 머리뿔은 크고 길게 발달되어 가슴 위로 솟아있다. 머리와 가슴은 검정색이며 딱지날개는 암적색을 띤다.

Inhabiting the regions of South America, this species has a pronotum horn split into two and bent in all directions while its cephalic horn, long and big, is developed up the thorax. Its head and thorax are black while its elytra (hard wings) are dark red.

[Distribution] Argentina

♂
볼리비아 산 44mm
(Bolivia. 1999.)

Heterogomphus

학 명 Scientific name	할르투스털장수풍뎅이 ***Heterogomphus hirtus***
채집국 Collected locality	🇧🇴 볼리비아 _ Bolivia
크 기 Size	♂ 35-50mm, ♀ 40mm

● 채집지
Collected Locality

● 분포지
Distribution

에콰도르 / 페루 / 콜롬비아 / 베네수엘라 / 볼리비아
Ecuador / Peru / Colombia / Venezuela / Bolivia

남아메리카 중북부에 서식한다. 이 종은 가슴뿔이 두텁고 넓게 두 갈래로 나뉘어져 전방향으로 굽어져 있으며 머리뿔은 길게 잘 발달되어 가슴 위로 솟아 있다. 몸빛은 암적색, 검정색을 띠며 황색의 털로 덮여있다.

It inhabits the central and northern regions of South America. It has a thick and wide pronotum horn bent in all directions while its cephalic long horn is well developed up the chest. Its dark red and black body is covered with yellow hairs.

[Distribution] Bolivia, Colombia, Ecuador, Venezuela, Peru

♂
프랑스 산 35mm
(Nimes, France. 2001. 7.)

Oryctes

학 명 Scientific name	나씨코르니스장수풍뎅이 *Oryctes nasicornis*
채집국 Collected locality	프랑스 _ France
크 기 Size	26-40mm

● 채집지 Collected Locality
● 분포지 Distribution

프랑스 France | 아프리카 북부 N. Africa | 터키 Turkey | 러시아남부 S. Russia | 인도 India | 부탄 Bhutan

유럽 전역과 아프리카 북부, 터키, 부탄까지 넓게 서식한다. 이 종은 가슴이 전방향으로 함몰되어 있어 가슴의 돌기가 발달한 것처럼 보인다. 머리뿔은 길고 안쪽으로 굽어져 있다. 유럽에서 발견되는 유일한 대형종으로 몸크기가 큰 것일수록 머리뿔이 발달하지 않는 경향을 보인다. 몸빛은 광택을 지니며 머리와 가슴은 검정색, 딱지날개는 암갈색을 띤다.

Its habitat is very wide from all across Europe to northern Africa, Turkey and Bhutan. As its thorax is dented in all directions, sometimes it looks like that the protrusions in their thorax are well developed. Its cephalic horn is long and bent inward. This species is the only major one discovered in Europe. The bigger one's body size, the less developed its cephalic horn. The head and thorax of its glossy body are black while the elytra (hard wings) are dark red.

[Distribution] Northern Europe, Northern Africa, Southern Russia, Turkey, India, Bhutan

♂
미얀마 남부 산 60mm
(S. Myanmar. 2002.)

Trichogomphus

학 명 Scientific name	마르타바니장수풍뎅이 *Trichogomphus martabani*
채집국 Collected locality	🇲🇲 미얀마 _ Myanmar
크 기 Size	♂ 35-62mm, ♀ 30-50mm

● 채집지
Collected Locality

● 분포지
Distribution

| 인도 | 미얀마 | 보르네오 북부 | 중국 |
| India | Myanmar | N. Borneo I. | China |

인디아, 인도차이나 반도, 보르네오 북부, 중국 전반에 걸쳐 넓게 서식한다. 이 종은 가슴중앙의 뿔이 위로 크게 발달하였으며 말단은 두 갈래로 나뉘어져 있다. 또한 가슴 양측에 돌기가 있으며 머리뿔은 크고 안으로 굽어져 있다. 수컷의 가슴은 사각형태로 되어있어 특징적이다. 몸빛은 검정색이며 점각이 있다.

This species has a wide habitat spanning from India, the Indochina Peninsula and Northern Borneo Island to all across China. It has a big and soaring horn from the center of its thorax, with a tip split into two. Some protrusions can be seen on both sides of its thorax while the big cephalic horn is bent inward. The male is unique in that it has a rectangular thorax. This species is black and has stipples.

[Distribution] India, Indochina Peninsula, Northern Borneo Island, China

♂
미국 아리조나 산 33mm
(U.S.A. Arizona. 2004.)

Xyloryctes

학 명 Scientific name	**자메이쎈씨스장수풍뎅이** ***Xyloryctes jamaicensis***
채집국 Collected locality	🇺🇸 미국 _ U.S.A.
크 기 Size	♂ 33 mm, ♀ 28mm

- 채집지
Collected Locality
- 분포지
Distribution

0°

미국(아리조나)　멕시코
U.S.A.(Arizona)　Mexico

북아메리카의 아리조나와 멕시코에 서식한다. 이 종은 가슴이 전방향으로 함몰되어 있고 가슴돌기는 발달하지 않았다. 머리뿔은 안쪽으로 굽어 있으며 암컷의 머리에는 작은 돌기가 있다. 몸빛은 검정색이며 암적색을 띠기도 한다.

This species inhabit Arizona, USA and Mexico. It has a thorax dented in all directions without protrusions. Its cephalic horn is bent inward while the female has a small projection on its head. It is mainly black and sometimes dark red.

[Distribution]　U.S.A.(Arizona) , Mexico

♂
전북 부안 산 65mm
(Mt. Byeonsan, Buan-gun, Jeollabuk-do, S. Korea. 1997. 7.)

Allomyrina

학 명 Scientific name	장수풍뎅이 *Allomyrina dichotomus*
채집국 Collected locality	🇰🇷 대한민국 _ Korea
크 기 Size	♂ 30-85mm, ♀ 43-55mm

- 채집지 Collected Locality
- 분포지 Distribution

중국 China | 인도차이나 반도 북부 N. Indochina pe. | 타이완 Taiwan | 대한민국 Korea | 일본 Japan

한국, 중국, 타이완, 일본, 인도차이나 북부의 산림지대에 서식한다. 이 종은 한반도산으로 수컷은 가슴위로 솟아 오른 두 갈래의 작은 뿔을 가지고 있으며 머리 부분에 앞으로 뻗어 힘차게 발달한 큰 뿔은 말단부가 네 갈래로 갈라져 있다. 몸빛은 검정색, 암적색을 띠며 광택을 지닌다.

This species, which originated from the Korean Peninsula, inhabits the forests in Korea, China, Taiwan, Japan and northern Indochina. The males have a couple of small horns rising up their thorax and another four-forked, big and hard one on the cephalic part. Its glossy body is black or dark red.

[Distribution] Korea, China, Taiwan, Japan, Northern of Indochina peninsular

♂
사라와크 산 32mm
(Sarawak I., N. Borneo. 2005. 2.)

Allomyrina

학 명 Scientific name	**프훼이훼리털장수풍뎅이** ***Allomyrina pfeifferi***
채집국 Collected locality	말레이시아 _ Malaysia
크 기 Size	♂ 27-40mm, ♀ 25-40mm

● 채집지 Collected Locality
● 분포지 Distribution

말레이 반도 / Malay Pe.   보르네오 / Borneo I.   술라웨시 섬 / Sulawesi I.   필리핀(민다나오 섬) / Philippines(Mindanao I.)

말레이 반도, 보르네오, 필리핀, 술라웨시 섬의 해발 1,500m이하의 산림에 서식한다. 수컷의 가슴위로 솟아 오른 뿔은 두 갈래로 갈라지지 않고 넓으며 머리 뿔은 말단이 두 갈래로 갈라져 안쪽으로 많이 굽어 있다. 몸빛은 암, 수 모두 쇠가 녹슨 것과 같은 색채를 띠고 있다.

This species inhabits the forests lower than 1,500m above sea level in the Malay Peninsula, Borneo Island, the Philippines and Sulawesi Island. The males` pronotum horns are wide and not forked into two while their cephalic horns are divided into two at the tip and bent much inward. Both the males and females have a rusty body color.

[Distribution] Malay Peninsula, Borneo Island, Philippines(Mindanao Island), Sulawesi Island

♂
타이 북부 산 103mm
(Chiang-Rai, N. Thailand. 2003. 7.)

Allomyrina

학 명 Scientific name	아틀라스장수풍뎅이 *Chalcosoma atlas*
채집국 Collected locality	 타이 _ Thailand
크 기 Size	♂ 45-110mm, ♀ 41-64mm

● 채집지
Collected Locality

● 분포지
Distribution

네팔 / 인도 북부 / 말레이 반도 / 수마트라섬 / 타이 / 술라웨시섬 / 필리핀
Nepal / NE. India / Malay Pe. / Sumatra I. / Thailand / Sulawesi I. / Philippines

인도 북동부로부터 인도차이나반도를 거쳐 술라웨시까지 넓게 서식하고 있다. 이 종의 수컷은 두 개의 가슴뿔이 양쪽으로 매우 길고 크게 잘 발달하여 있다. 머리뿔도 길게 잘 발달하여 안으로 굽어있다. 이 종은 태국북부에 서식하는 아틀라스장수풍뎅이로, 몸빛은 검정색으로 보이나 금속성의 색채와 광택을 가지고 있으며 산지에 따라 금속성색채가 다르게 나타난다.

This species has an expansive habitat across India, Nepal and the entire Southeast Asian region to the Philippines. The males have a pair of long and big thorox horns, well developed in both directions. Their long and well-developed cephalic horns are bent inward. This species, the atlas rhinoceros beetle inhabiting northern Thailand, has a metallic-colored and glossy body though it looks black at the first glance. The metallic color shows some variations depending on where the habitat is.

[Distribution] NE. India, Nepal, Indochina Peninsula, Malay Peninsula, Philippines, Sumatra Island, Borneo Island, Sulawesi Island.

♂
술라웨시 산 95mm
(Sulawesi, Indonesia. 2002. 6.)

Chalcosoma

학 명 Scientific name	아틀라스장수풍뎅이 *Chalcosoma atlas*
채집국 Collected locality	인도네시아 _ Indonesia
크 기 Size	♂ 45-108mm, ♀ 41-64mm

● 채집지 Collected Locality
● 분포지 Distribution

술라웨시 섬
Sulawesi I.

인도네시아 술라웨시섬 산으로 뿔의 형태가 다양하다. "*Chalcosoma*"는 "금속적인 광택을 하고 있다"는 뜻으로 몸빛은 검정색으로 보이나 황색, 녹색, 청색의 금속성 색채가 다양하다. 성충의 수명은 약 2개월 정도로 짧다. 번데기에서 성충으로 용환된 젊은 개체일수록 광택이 강하고 수명을 다할수록 개체의 광택은 약해진다. 대형 개체에서 장각과 단각이 같이 나타난다. 소형 개체 또한, 장각과 단각이 같이 나타난다. 이는 서식 지역의 표고와 온도, 먹이 형태에 따라 관계되어지며 표고가 높을수록(해발 1,400m) 대형의 개체가 나타난다.

Originated from Sulawesi Island, Indonesia, this rhinoceros beetle has a variety of horn shapes. The word '*Chalcosoma*' means metallic gloss. Its metallic body, though it appears black, actually consists of various colors yellow, green, blue, etc. The lifespan of its adult insects is as short as two months. The younger adult insects are glossier while those older have less gloss. Its large individuals, like their smaller counterparts, show both long and short horns. This is related to the altitude and temperature of their habitats as well as what they feed on. The higher the altitude (1,400m above sea level) is, the larger the specimens are likely to be.

[Distribution] Sulawesi Island

♂
필리핀 민다나오 산 95mm
(Mt. Apo, Mindanao I., Philippines. 2004. 8.)

Chalcosoma

학 명 Scientific name	아틀라스장수풍뎅이 *Chalcosoma atlas*
채집국 Collected locality	필리핀 _ Philippines
크 기 Size	♂ 67-110mm, ♀ 41-64mm

필리핀산으로 뿔의 형태가 다양하다. 특히 말레이 반도의 개체가 대형(100mm)에서부터 소형(50mm이하)에 이르기까지 다양하게 나타나는데 반해 필리핀의 민다나오 섬 아포산(Mt. Apo, 2,952m)의 개체들은 90mm의 대형 개체들이 주류를 이루고 있다. 몸빛은 검정색으로 황색, 붉은색의 금속성 색채와 광택을 지닌다. 중형 및 소형 개체에 따라서 머리뿔이 세 갈래로 나뉘는 특징이 있다.

Originated from the Philippines, this species has diverse horn shapes. Especially, those living in the Malay Island show a wide variety of body sizes from 100m-long to 50mm-long or smaller. On the contrary, those inhabiting Mindanao Island and Mt. Apo (2,952m above sea level), Philippines, are relatively larger; 90mm-long on average. Their black body shows some metallic yellow or red hue and gloss. Some middle- and small-sized individuals have three-forked cephalic horns.

[Distribution] Philippines

♂
수마트라 서부 산 105mm
(Landai, W. Sumatra, Indonesia. 2005. 5.)

Chalcosoma

학 명 Scientific name	키론장수풍뎅이 ***Chalcosoma chiron***
채집국 Collected locality	인도네시아 _ Indonesia
크 기 Size	♂ 55-127mm, ♀ 50-85mm

- 채집지
Collected Locality
- 분포지
Distribution

수마트라 섬 Sumatra I.   자바 섬 Java I.

인도네시아의 수마트라와 자바 섬에 서식하는 키론장수풍뎅이이다. 수컷의 머리뿔 상부의 돌기는 말레이 반도산에 비하여 대체로 매끄럽고 가늘다. 몸빛은 검정색 바탕에 황청색의 금속성 광택을 지니고 있다.

This chiron rhinoceros beetle species inhabits Sumatra & Java Island, Indonesia. The protrusions on the upper part of males` cephalic horns are softer and slenderer than those of their counterparts living in the Malay Peninsula. This species has a black body with a metallic yellowish-blue gloss.

[Distribution] Sumatra Island, Java Island

♂
보르네오 북부 산 91mm
(Mt. Kota Kinabalu, 1,400m. Sabah, Borneo, E. Malaysia. 2004. 5.)

Chalcosoma

학 명 Scientific name	**몰렌캄피장수풍뎅이** *Chalcosoma mollenkampi*
채집국 Collected locality	말레이시아 _ Malaysia
크 기 Size	♂ 50-112mm, ♀ 50-60mm

- 채집지 Collected Locality
- 분포지 Distribution

보르네오 북부
N.Borneo I.

보르네오 섬 북부에 서식한다. *C. atlas, C. chiron*에 비하여 작은 가슴에 직선상으로 발달된 가슴뿔로 쉽게 구별이 가능하다. 몸빛은 검정색 바탕에 황녹색, 암갈색의 금속성 광택을 지니고 있다. 가슴은 매끈하지 않고 광택이 약하다. 이 종은 같은 보르네오 섬에 서식하는 *C. atlas*가 저지대에서부터 고지대에 이르기까지 넓게 서식하는 반면 동남아의 최고봉인 코타키나발루산 (해발 4,095m)의 표고 1,000m이상의 지점등 고지에 한정되어 서식한다. 개체의 크기가 작아짐에 따라 뿔크기도 같이 작아 지는 특성을 지녔다.

Inhabiting nothern Borneo Island, this species is easily distinguished from *C. atlas* and *C. chiron* due to its smaller thorax and linear pronotum horns on it. This species has a black body with some metallic yellowish-green or dark-brown gloss. The thorax, not so soft, tends to be less glossy. Though cohabiting in Borneo Island with *C. atlas*, this species` habitat is limited to the area 1,000m or higher in Mt. Kota Kinabalu (4,095m above sea level), the highest peak in Southeast Asia, unlike *C. atlas*, which has an expansive habitat in terms of altitude. Smaller individuals of this species characteristically have smaller horns and vice versa.

[Distribution] Nothern Borneo Island

♂
콩고 산 76mm
(Congo. 1997.)

Augsoma

학 명 Scientific name	켄타우르스장수풍뎅이 ***Augsoma centaurus***
채집국 Collected locality	콩고 _ Congo
크 기 Size	♂ 40-90mm, ♀ 40-60mm

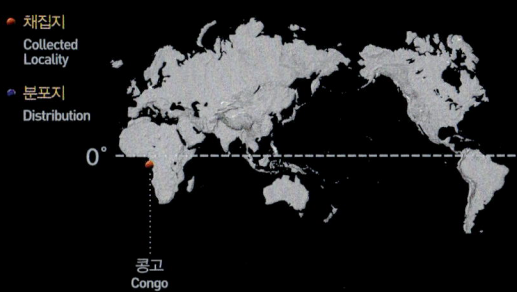

- 채집지 Collected Locality
- 분포지 Distribution

콩고
Congo

중앙아프리카 서부에 서식한다. 한 개의 긴 가슴뿔과 양쪽에 돌기가 발달해 있다. 머리뿔 안쪽의 상단부는 마치 통조림 따개처럼 생겼다. 머리와 가슴은 검정색이며 딱지날개는 암적색을 띠고 모두 강한 광택을 지니고 있다.

Inhabiting central and west Africa, this species has one long pronotum horn with some protrusions on both its sides. The upper inside of their cephalic horn looks like a can opener. Their cephalic part and thorax are black while the elytra (hard wings) are dark-red, all of which are very glossy.

[Distribution] Central and West Africa

♂
미국 아리조나 산 52mm
(Anizona, U.S.A.)

Dynastes

학 명 Scientific name	그란티장수풍뎅이 ***Dynastes granti***
채집국 Collected locality	🇺🇸 미국 _ U.S.A.
크 기 Size	♂ 35-85mm, ♀ 37-50mm

● 채집지 Collected Locality
● 분포지 Distribution

미국(아리조나)
U.S.A.(Anizona)

멕시코와 근접하는 북아메리카 남부 아리조나에 서식한다. 북아메리카에 서식하는 *Dynastes*속 장수풍뎅이는 3종으로 *D.tityus*와 *D.hyllus*와 더불어 3종이 있다. *Dynastes*속의 장수풍뎅이는 가슴뿔과 머리뿔은 한 개씩으로 전방향으로 길게 발달하였고 그 밑에 솔과 같은 황색의 강모가 나있는 특성을 지닌다. 딱지날개는 불규칙적인 짙은갈색의 점무늬를 가지고 개체별로 그 무늬가 다르다.

They inhabit Arizona and the southern part of North America, close to Mexico. The *Dynastes* genus inhabiting North America is divided into three, including *D.tityus* and *D.hyllus*. Both of them are relatively smaller than the other rhinoceros beetles in the same genus. Compared to *D.tityus*, they look brighter and have a light yellowish-grey color while their elytra (hard wings) have some irregular dark-brown speckles, whose shapes are individually different.

[Distribution] U.S.A.(Anizona)

♂
미국 남동부 산 55mm
(SE. U.S.A.. 2006. 7.)

Dynastes

학 명 Scientific name	**티티우스장수풍뎅이** ***Dynastes tityus***
채집국 Collected locality	🇺🇸 미국 _ U.S.A.
크 기 Size	♂ 40-70mm, ♀ 40-55mm

- 채집지 Collected Locality
- 분포지 Distribution

미국 남동부
SE. U.S.A.

북아메리카 동부에 넓게 서식한다. *D. granti*와 함께 북아메리카에 서식하는 중소형의 장수풍뎅이이다. 몸빛은 짙은 황색으로 말단부들은 검정색이다. 본 표본은 기본형과 블랙형의 변이형으로 판단된다. 이 종의 몸빛은 회색, 회청색, 검정색 등 색깔의 변이가 많고 딱지날개의 점무늬 또한 변화가 많다.

This rhinoceros beetle species has a wide habitat in eastern part of North America. This is one of the two middle- and small-sized rhinoceros beetle species inhabiting North America (the other is *D. granti*). Its dark-yellow body has black tips. This specimen is considered a variation from its basic type and black type. The body color of this species shows a lot of variations from grey, grayish-blue to black, like the spotted pattern on its elytra (hard wings).

[Distribution] Eastern and Southern U.S.A.

↑
에콰도르 산 143mm
(Napo, Cosanga, Ecuador. 2005. 3.)

Dynastes

학 명 Scientific name	**헤라클레스리치장수풍뎅이** ***Dynastes hercules lichyi***
채집국 Collected locality	🇪🇨 에콰도르 _ Ecuador
크 기 Size	♂ 60-170mm, ♀ 23.3-52mm

- 채집지 Collected Locality
- 분포지 Distribution

에콰도르 / Ecuador
콜롬비아 / Colombia
베네수엘라 및 인근 섬 / Venezuela and near by Islan

남아메리카 북부의 산림 1,000m 내외의 고지에 서식하고 있다. 머리와 가슴은 검정색으로 광택을 띠고 있다. 다른 *D. hercules* 에 비하여 미절판에 황색의 강모가 길게 밀집되어있다. 검은 반점은 불규칙하며 개체별로 다르다. *D. hercules* 대형종은 가슴뿔이 체장의 반을 차지할 정도로 길게 발달하였다. 뿔의 크기는 서식지의 온도와 표고, 먹이 등에 따라 다르게 나타난다. 일반적으로 우기 초기에 출현한다.

They inhabit the plateau in some northern parts of South American forests, approximately 1,000m-high. Their cephalic part and thorax are black and glossy. Compared with the other *D. hercules* individuals, the back part of its last abdominal segment is cpygidium denser with yellow bristles. The black speckles show no regularity or individual similarity. Larger *D. hercules* individuals` pronotum horns are so long that they are about a half of their body length. The altitude and temperature of their habitat and their food decide the size of the horns. They, in general, appear during an early wet season.

[Distribution] Ecuador, Colombia, Venezuela and near by islands

♂
베네수엘라 산 130mm
(Venezuela)

Dynastes

학 명 Scientific name	헤라클레스리치장수풍뎅이 *Dynastes hercules lichyi*
채집국 Collected locality	에콰도르 _ Ecuador
크 기 Size	♂ 60-170mm, ♀ 23.3-52mm

- 채집지 Collected Locality
- 분포지 Distribution

에콰도르 Ecuador 콜롬비아 Colombia 베네수엘라 및 인근 섬 Venezuela and near by Islands

남아메리카 북부의 산림 1,000m 내외의 고지에 서식하고 있다. 머리와 가슴은 검정색으로 광택을 띠고 있다. 다른 *D. hercules* 에 비하여 미절판에 황색의 강모가 길게 밀집되어있다. 검은 반점은 불규칙하며 개체별로 다르다. *D. hercules* 대형종은 가슴뿔이 체장의 반을 차지할 정도로 길게 발달하였다. 뿔의 크기는 서식지의 온도와 표고, 먹이 등에 따라 다르게 나타난다. 일반적으로 우기 초기에 출현한다.

They inhabit the plateau in some northern parts of South American forests, approximately 1,000m-high. Their cephalic part and thorax are black and glossy. Compared with the other *D. hercules* individuals, the back part of its last abdominal segment is cpygidium denser with yellow bristles. The black speckles show no regularity or individual similarity. Larger *D. hercules* individuals` pronotum horns are so long that they are about a half of their body length. The altitude and temperature of their habitat and their food decide the size of the horns. They, in general, appear during an early wet season.

[Distribution] Ecuador, Colombia, Venezuela and near by islands

♂
베네수엘라 산 130mm
(Venezuela)

Dynastes

학 명 Scientific name	넵튠장수풍뎅이 ***Dynastes neptunus***
채집국 Collected locality	베네수엘라 _ Venezuela
크 기 Size	♂ 50-155mm, ♀ 64-70mm

- 채집지 Collected Locality
- 분포지 Distribution

콜롬비아 Colombia · 에콰도르 Ecuador · 페루 북서부 NW. Peru · 베네수엘라 북서부 NW. Venezuela

남아메리카 북서부에 서식하고 있다. *D. hercules*종에 비하여 가슴뿔과 머리뿔이 더 길게 발달하여 있다. 특히 가슴 전방으로 두 개의 긴 돌기가 유독히 발달해있는 특징을 갖는다. 머리와 가슴은 검정색으로 강한 광택을 지니고 있다. 안데스 산맥의 표고 1,000m이상에 서식하고 있으며 활동시기는 새벽 3시부터이며 해가 진 후 활동하는 *D. hercules*와 서식지에서의 활동시간이 다르다. 배면 전체에 짙은 황색의 털이 나있다.

This species, which inhabits the northwestern part of South America, has cephalic and pronotum horns longer than those of *D. hercules*. The two long and highly developed protrusions on its thorax are especially peculiar. It has a black and highly glossy cephalic part and thorax. They live at an altitude of 1,000m or higher in the Andes. They are active from 3 a.m. unlike *D. hercules* which becomes active after sunset. They have dense, dark-yellow hair throughout their backs.

[Distribution] Ecuador, Colombia, Northwestern Venezuela, Northwestern Peru

↑
이리안 자야 서부 산 55mm
(W. Irian jaya, Indonesia. 1998.)

Eupatoraus

학 명 Scientific name	벡카리삼각뿔장수풍뎅이 ***Eupatorus beccarii***
채집국 Collected locality	인도네시아 _ Indonesia
크 기 Size	♂ 26-55mm, ♀ 23-42mm

● 채집지
Collected Locality

● 분포지
Distribution

뉴기니아 섬(이리안 자야)
New Guinea I.(Irian jaya)

뉴기니 섬에 서식한다. 두 개의 가슴뿔이 있으며 전방향으로 발달한 대형의 개체에서는 안쪽으로 톱과 같은 돌기가 이어져 있다. 머리뿔도 길게 잘 발달하여 안으로 굽어져 있다. 머리와 가슴은 광택을 지닌 검정색이며 딱지날개는 밝은 황갈색, 황적색으로 미약한 광택을 지닌다.

Inhabiting New Guinea, they have a pair of pronotum horns which develop in the forward direction. Larger individuals have some jagged protrusions pointed in. Their cephalic horns are also well-developed and bent inward. Their cephalic part and thorax are black and glossy while the light yellowish-brown or yellowish-red elytra (hard wings) are less glossy.

[Distribution] New Guinea Island

↑
타이 북부 산 55mm
(Chiang-Mai, N. Thailand. 2002.)

Eupatoraus

학 명 Scientific name	비르마니쿠스오각뿔장수풍뎅이 ***Eupatorus birmanicus***
채집국 Collected locality	타이 _ Thailand
크 기 Size	↕ 45-58mm

○ 채집지
  Collected
  Locality

○ 분포지
  Distribution

미얀마 / 타이 서북부
Myanmar / NW. Thailand

미얀마와 타이 서북부에 서식한다. 다섯 개의 뿔을 가지고 있다. 중앙부 두 개의 뿔은 마치 토끼 귀처럼 솟아 있으며 비교적 귀여운 외형의 장수풍뎅이이다. 머리뿔은 가늘고 길게 발달하여 안쪽으로 굽어 있다. 나머지 두 개의 뿔은 가슴전방 양쪽으로 돌기처럼 작게 발달하였다. 머리와 가슴은 검정색이며 딱지날개는 암갈색으로 모두 미약한 광택을 지니고 있다.

Having five horns, they inhabit Myanmar and the northwestern part of Thailand. The two horns in the center look much like rabbit ears while the cephalic one, long and slender, is bent inward. The other two horns are rather small, like a pair of protrusions developing from the thorax. The cephalic part and thorax are black while the elytra (hard wings) are dark-brown, all of which are slightly glossy.

[Distribution] Myanmar, Northwestern Thailand

↑
타이 북부산 80mm
(Chiang-Rig, N. Thailand. 2001. 7.)

Eupatoraus

학 명 Scientific name	그라실리코르니스오각뿔장수풍뎅이 *Eupatoraus gracilicornis*
채집국 Collected locality	타이 _ Thailand
크 기 Size	♂ 50-80mm, ♀ 40-50mm

- 채집지 Collected Locality
- 분포지 Distribution

인도북동부 NE. India　중국 남부 S. china　미안마 Myanmar　타이 Thailand　베트남 Vietnam　라오스 Laos

인도 북동부, 인도차이나 북부, 중국 남부에 서식한다. 다섯 개의 뿔을 갖는 장수풍뎅이 중 가장 대표종이다. 가슴에 네 개의 뿔과 머리에 길게 발달한 긴 뿔은 과거 인도차이나 반도의 무장된 전투 코끼리를 연상케 한다. 표고 1,000m이상의 죽림에 서식하며 성충은 죽순을 상처 내어 즙을 핥아 먹는다. 소형의 개체에서 발달하지 못한 뿔은 *E. sukkiti*와 혼동될 수 있다.

This species, inhabiting India, southern China and northern Indochina, is typical of the rhinoceros beetles having five horns. The four horns on the thorax and the other developed long on the cephalic part gives it the look of an Indochinese war elephant of the past. It lives in bamboo forests at an altitude of 1,000m or higher. Its adults scratch bamboo shoots and lick the juice. Smaller individuals, whose horns are relatively underdeveloped, look much like *E. sukkiti*.

[Distribution] Northestern India, Sounthern China, Myanmar, Thailand, Laos, Vietnam

↑
타이 북부산 58mm
(Chiang-Rai, N. Thailand. 2004. 7.)

Eupatorus

학 명 Scientific name	씨아멘시스오각뿔장수풍뎅이 ***Eupatorus siamensis***
채집국 Collected locality	타이 _ Thailand
크 기 Size	♂ 50-70mm, ♀ 40-50mm

● 채집지
Collected Locality

● 분포지
Distribution

타이 / 라오스 / 베트남
Thailand / Laos / Vietnam

인도차이나 북부에 서식한다. *E. gracilicornis*와 비교하여 가슴뿔은 짧지만 두텁고 넓게 발달하였다. 가슴 위의 두 개의 뿔은 소뿔을 닮았다. 소형개체는 뿔의 발달이 좋지 못하다. 머리와 가슴은 검정색이며 딱지날개는 암갈색의 미약한 광택을 지닌다.

This species, which inhabits the northern part of Indochina, has pronotum horns shorter but thicker and wider than those of *E. gracilicornis*. The two horns on the thorax are like those of a cow. Smaller specimens` horns are less developed. The cephalic part and thorax are black while the elytra (hard wings) are dark-brown and slightly glossy.

[Distribution] Thailand, Laos, Vietnam

♂
미얀마 북부 산 59mm
(Kachin, N. Myanmar. 2003. 8.)

Eupatorus

학 명 Scientific name	수키티오각뿔장수풍뎅이 ***Eupatorus sukkiti***	
채집국 Collected locality		미얀마 _ Mayanmar
크 기 Size	↕ 45-63mm	

- 채집지 Collected Locality
- 분포지 Distribution

미얀마 북부
N. Myanmar

미얀마 북부에 서식한다. *E. gracilicornis*의 소형종과 비교하여 가슴 중앙 두 개의 뿔은 좀 더 위로 솟아있다. 가슴양쪽의 뿔은 위로 굽어있지 않고 전방으로 짧게 발달해 있다. 머리와 가슴은 광택을 지닌 검정색이며 딱지날개는 짙은 암갈색의 미약한 광택을 지닌다.

The two horns in the thorax center of this species inhabiting northern Myanmar point upward higher than those of smaller *E. gracilicornis* individuals. The horns on both sides of the thorax, not bent upward, are developed short and in the forward direction. The cephalic part and thorax are black and glossy while the elytra (hard wings) are dark-brown and slightly glossy.

[Distribution] Northern Myanmar

♂
페루 산 50mm
(Peru. 2002.)

Golofa

학 명 Scientific name	크라비게르앞장다리장수풍뎅이 ***Golofa claviger***
채집국 Collected locality	🇵🇪 페루 _ Peru
크 기 Size	↕ 35-60mm

- 채집지 Collected Locality
- 분포지 Distribution

파나마 Panama | 콜롬비아 Colombia | 가이아나 Guyanan | 페루 Peru | 볼리비아 Bolivia | 브라질 Brazil

중앙아메리카 남부와 남아메리카 중북부에 서식한다. 위로 솟은 가슴뿔의 상부는 크고 넓적하게 세 갈래로 나뉘어져 있는 매우 큰 특징을 가지고 있다. 가슴뿔 안쪽으로 황색의 강모를 가지고 있다. 전체적으로 부드러운 광택을 지니고 있다.

They inhabit southern Central America and the middle and northern parts of South America. The upper part of the pronotum horn pointing upward is wide and triple-forked. Inside the pronotum horn are yellow bristles. Generally, their body is glossy and soft.

[Distribution] **Panama, Guyana, Colombia, Brazil, Peru, Bolivia**

♂
멕시코산 50mm
(Mexico. 2003.)

Golofa

학 명 Scientific name	피자로앞장다리장수풍뎅이 *Golofa pizarro*
채집국 Collected locality	🇲🇽 멕시코 _ Mexico
크 기 Size	↕ 30-52mm

● 채집지
Collected Locality

● 분포지
Distribution

0°

| 멕시코 | 엘살바도르 | 온두라스 | 과테말라 | 니카라과 | 코스타리카 | 파나마 |
| Mexico | El Salvador | Honduras | Guatemala | Nicaragua | Costa Rica | Panama |

중앙아메리카와 남아메리카 중북부에 서식한다. *G. claviger*와 비교하여 위로 솟은 가슴뿔의 상부는 크고 넓적하게 삼각형으로 되어 있다. 가슴뿔 안쪽으로 황색의 강모를 가지고 있다. 머리뿔 안쪽으로 톱날과 같은 돌기가 있다. 전체적으로 부드러운 광택을 지니고 있다.

They inhabit Central America and the middle and northern parts of South America. Compared with *G. claviger*, the upper part of the pronotum horn pointing upward is wide and triangular. Inside the pronotum horn are yellow bristles while some protrusions like saw blades are inside the cephalic horn. Generally, their body is glossy and soft.

[Distribution] Mexico, Guatemala, Honduras, Nicaragua, El Salvador, Costa Rica, Panama

♂
페루 산 50mm
(Peru. 2004.)

Golofa

학 명 Scientific name	에아쿠스앞장다리장수풍뎅이 *Golofa eacus*
채집국 Collected locality	페루 _ Peru
크 기 Size	♂ 30-50mm, ♀ 30-40mm,

- 채집지 Collected Locality
- 분포지 Distribution

베네수엘라 Venezuela, 에콰도르 Ecuador, 콜롬비아 Colombia, 페루 Peru, 볼리비아 Bolivia, 아르헨티나 Argentina, 브라질 Brazil

남아메리카 전반에 걸쳐 넓게 서식한다. 가늘고 길게 발달한 위로 솟으며 발달한 머리뿔과 가슴뿔이 특징적이다. 가슴뿔 안쪽으로 솔과 같은 황색의 강모가 나있고 머리뿔은 상단부에 거친 톱날 같은 돌기가 있다. 대나무를 주식으로 하는 이 종은 앞다리가 대나무 줄기를 잘 올라 갈 수 있도록 발달되었으며 공격과 방어 및 수컷끼리의 싸움에 이용된다.

Having a wide habitat across South America, they are characterized by their long and slender cephalic and pronotum horns which point upward. Inside the pronotum horns are brush-like yellow bristles while the upper part of the cephalic horns has some protrusions like saw blades. Their forelegs are suitable for climbing up bamboo trees, which they mainly feed on. The males also use their forelegs to fight or protect themselves from one another.

[Distribution] Venezuela, Colombia, Ecuador, Peru, Brazil, Bolivia, Argentina

♂
볼리비아 산 39mm
(Bolivia. 2004.)

Golofa

학 명 Scientific name	**펠라곤앞장다리장수풍뎅이** *Golofa pelagon*
채집국 Collected locality	🇧🇴 볼리비아 _ Bolivia
크 기 Size	♂ 27-45mm, ♀ 23-33mm,

- 채집지 Collected Locality
- 분포지 Distribution

0°

콜롬비아 Colombia 볼리비아 Bolivia 아르헨티나 Argentina 브라질 Brazil

남아메리카 전반에 걸쳐 넓게 서식한다. 가늘고 길게 위로 솟으며 발달한 머리뿔과 가슴뿔이 특징적이다. 가슴뿔 안쪽으로 솔과 같은 황색의 강모가 나왔고, *G. eacus*에 비해 머리뿔의 상단부에 거친 톱날 같은 돌기는 없다. 대나무를 주식으로 하는 이 종은 앞다리가 대나무 줄기를 잘 올라 갈 수 있도록 발달되었으며 공격과 방어 및 수컷끼리의 싸움에 이용된다.

They, having a wide habitat across South America, are characterized by their long and slender cephalic and pronotum horns pointing upward. Inside the pronotum horns are brush-like yellow bristles while the upper part of the cephalic horns has some protrusions like saw blades. Their forelegs are suitable for climbing up bamboo trees, which they mainly feed on. The males also use their forelegs to fight or protect themselves from one another.

[Distribution] Colombia, Brazil, Bolivia, Argentina

♂
페루 산 116mm
(Pucacuru, Loreto, Peru. 2006. 12.)

Megasoma

학 명 Scientific name	악테온도깨비뿔장수풍뎅이 ***Megasoma actaeon***
채집국 Collected locality	🇵🇪 페루 _ Peru
크 기 Size	♂ 50-135mm, ♀ 70-77mm

● 채집지 Collected Locality
● 분포지 Distribution

에콰도르	콜롬비아	페루 북부	베네수엘라	기아나
Ecuador	Colombia	N. Peru	Venezuela	Guiana

남아메리카 북부에 서식한다. 양쪽의 가슴뿔은 원뿔형으로 두껍고 강하게 발달하여 전방부로 뻗어있다. 머리뿔의 말단부는 두 갈래로 나뉘어져 있으며 기저부에 위로 솟은 하나의 큰 돌기가 발달해 있다. 온몸이 검정색이며 부드러운 광택을 지니고 있다. *M. e. elephas*에 버금가는 크기와 체중을 지니고 있다.

They inhabit the northern part of South America. Their cone-shaped pronotum horns on both side of the thorax are thick and hard, and pointed forward. The cephalic horns` tips are double-forked with a big protrusion rising upward from the base. Their body, black and softly glossy, is as heavy and big as that of *M. e. elephas*.

[Distribution] Venezuela, Colombia, Ecuador, Guiana, Northern Peru

♂
페루 산 105mm
(Loreto Eden, Satipo, Peru. 2008. 5.)

Megasoma

학 명 Scientific name	마르스도깨비뿔장수풍뎅이 *Megasoma mars*
채집국 Collected locality	페루 _ Peru
크 기 Size	♂ 70-140mm, ♀ 70-76mm

- 채집지 Collected Locality
- 분포지 Distribution

콜롬비아 / 페루 / 가이아나 / 브라질
Colombia / Peru / Guyanan / Brazil

남아메리카 중북부에 서식한다. *M.actaeon*와 비교하여 양쪽의 가슴뿔은 가늘게 발달하여 바깥으로 뻗어있다. 머리뿔의 말단부는 두 갈래로 나뉘어져 있으며 기저부에 위로 솟은 하나의 큰 돌기가 발달해 있다. 온몸이 검정색으로 강한 광택을 지니고 있다.

They inhabit the middle and northern parts of South America. Compared with *M.actaeon*, their bidirectional pronotum horns are slenderer and developed outward. The cephalic horns' tips are two-forked with a big protrusion rising upward from the base. Their body is black and highly glossy.

[Distribution] Brazil, Guyana, Peru, Colombia

♂
멕시코 산 113mm
(San Pedro, Soteapan 500m, Mexico. 2006. 9.)

Megasoma

학 명 Scientific name	엘라파스코끼리장수풍뎅이 ***Megasoma elephas elephas***
채집국 Collected locality	멕시코 _ Mexico
크 기 Size	♂ 50-120mm, ♀ 50-70mm

중앙아메리카와 남아메리카 북부에 서식한다. 코끼리의 코처럼 앞으로 길게 뻗은 머리뿔과 가슴 양쪽으로 돌출된 뿔은 마치 코끼리의 상아를 보는 듯하여 코끼리장수풍뎅이라 부른다. 그 이름에 걸맞게 체중도 가장 무거운 장수풍뎅이이다. 온몸에는 황색의 미모가 나있다.

It inhabits Central America and the northern part of South America. Its long cephalic horns, which develop in the forward direction, and pronotum horns protruded in both directions of the thorax resemble an elephant's nose and ivory, which gave it the name, 'elephant beetle.' Its weight, matches its name well. One of the heaviest rhinoceros beetle species in fact, this species is covered with yellow fine hair.

[Distribution] Southern Mexico, Northern Colombia, Northern Venezuela, Caribbean Is.

♂
브라질 산 68mm
(Amazon, Brazil. 2005. 1.)

Megasoma

학 명 Scientific name	기아스코끼리장수풍뎅이 *Megasoma gyas rumbucheri*
채집국 Collected locality	브라질 _ Brazil
크 기 Size	65-115mm

- 채집지 Collected Locality
- 분포지 Distribution

가이아나 Guyana 브라질 북부 N. Brazil

남아메리카 북동부에 서식한다. *M.e.elephas*와 비교하여 크기가 작으며 머리뿔이 매우 짧아 아기코끼리와 닮아 있다. 중앙부 가슴 뿔은 오히려 발달하여 전방부로 향하여 있다. 온몸에는 황색의 미모가 나있다.

This species inhabits the northeastern part of South America. It is smaller than *M.e.elephas* and its very short cephalic horns are somewhat like the nose of a young elephant. On the contrary, the central pronotum horn is well-developed in the forward direction. Its entire body is covered with fine yellow hair.

[Distribution] Guyana, Northern Brazil

♂
아르헨티나 산 35mm
(Argentina. 2005.)

Megasoma

학 명 Scientific name	조에르겐세니코끼리장수풍뎅이 *Megasoma joergeseni joergenseni*
채집국 Collected locality	아르헨티나 _ Argentina
크 기 Size	♂ 35mm, ♀ 31mm

- 채집지 Collected Locality
- 분포지 Distribution

파라과이 Paraguay   아르헨티나 Argentina

남아메리카의 파라과이와 아르헨티나에 서식한다. *M. e. elephas*와 비교하여 크기가 매우 작으며 머리뿔은 *M. g. rumbucheri*보다도 더 짧다. 중앙부 가슴뿔은 두텁고 뭉뚝하게 발달하여 전방부로 향하여 있다. 온몸에는 보다 긴 황색의 미모가 나있다.

This species inhabits Paraguay and Argentina. It is much smaller than *M. e. elephas* and its cephalic horn is even shorter than that of *M. g. rumbucheri*. The central pronotum horn, thick and stumpy, is pointed forward. Its entire body is covered with yellow, fine and relatively long hair.

[Distribution] Paraguay, Argentina

♂
멕시코 산 35mm
(Baja California Sur, Todos Santos, Mexico. 2006. 9.)

Megasoma

학 명 Scientific name	**테르시테스코끼리장수풍뎅이** ***Megasoma thersites***
채집국 Collected locality	🇲🇽 멕시코 _ Mexico
크 기 Size	♂ 27-45mm

● 채집지 Collected Locality
● 분포지 Distribution

멕시코 바하 캘리포니아 남부
Mexico(Baja California)

중앙아메리카 멕시코 바하 캘리포니아 남부의 한정된 지역에 서식한다. *M. e. elephas*와 비교하여 크기가 매우 작으며 머리뿔은 *M. g. rumbucheri*보다도 더 짧다. 중앙부 가슴뿔은 가늘고 말단부는 두 갈래로 나뉘어져 전방부로 향하여 있다. 온몸에는 *M. j. joergenseni*보다 긴 황색의 장모가 나있다.

This species` habitat is southern Baja California areas in Mexico. It is much smaller than *M. e. elephas* and its cephalic horn is even shorter than that of *M. g. rumbucheri*. The thin central pronotum horn is double-forked at its tip pointing forward. The yellow and long hair covering its body is longer than that of *M. g. rumbucheri*.

[Distribution] Mexico(Baja California)

플로네스 산 70mm
(Flores I., Indonesia. 2004. 4.)

Xylotrupes

학 명 Scientific name	기데온플로넨시스장수풍뎅이 ***Xylotrupes gideon florensis***
채집국 Collected locality	인도네시아 _ Indonesia
크 기 Size	↕ 40-75mm

● 채집지
Collected Locality

● 분포지
Distribution

인도네시아(플로네스 섬)
Indonesia(Flores I.)

인도네시아의 플로네스(Flores) 섬에 서식한다. 이 아종은 가슴뿔 기저부 양쪽에 세 갈래로 나뉘어진 돌기를 가지고 있는 큰 특징이 있다. 머리와 가슴은 검정색이며 딱지날개는 암적색을 띤다. 부드러운 광택을 지니고 있다.

This subspecies, inhabiting Flores Island, Indonesia, is characterized by its protrusions divided into three on both sides of the pronotum horn base. The cephalic part and thorax are black while the elytra (hard wings) are dark-red. Its body is softly glossy.

[Distribution] Indonesia(Flores Island)

♂
필리핀 민다나오 산 45mm
(Mindanao I., Philippines. 2004.)

Xylotrupes

학 명 Scientific name	프베센스털장수풍뎅이 *Xylotrupes pubescens*
채집국 Collected locality	필리핀 _ Philippines
크 기 Size	♂ 35-55mm

● 채집지 Collected Locality
● 분포지 Distribution

필리핀 남부(민다나오 섬)
S. Philippines(Mindanao I.)

필리핀 남부에 서식한다. 이 종은 *Xylotrupes*속 중 유일하게 황색의 미모가 나있다. 몸크기에 비하여 머리와 가슴뿔은 길게 발달하지 않았다. 몸빛은 암갈색으로 부드러운 광택을 지니고 있다.

This species, inhabiting the southern Philippines, is the only species among the *Xylotrupes* genus which has yellow fine hair. In proportion to the body size, the thorax and cephalic horns are not much developed. Its dark-red body is softly glossy

[Distribution] Southern Philippines(Mindanao Island)

♂
전북 부안 산 22mm
(Mt. Byeonsan, Buan-gun, Jeollabuk-do, S. Korea. 2004. 8.)

Pentodon

학 명 Scientific name	**외뿔장수풍뎅이** *Pentodon quadridens*
채집국 Collected locality	🇰🇷 대한민국 _ Korea
크 기 Size	♂ 18-25mm, ♀ 18-20mm

● 채집지
  Collected
  Locality

● 분포지
  Distribution

중국 China, 타이완 Taiwan, 대한민국 Korea, 일본 Japan

동남아시아 및 극동아시아에 서식한다. 이 종은 암, 수 모두 머리에 작은 뿔을 가지고 있으며 수컷의 가슴중앙부는 움푹 들어가 있는 특징을 가진다. 몸빛은 검정색이고 세로줄의 점각이 뚜렷하며 약한 광택을 지닌다. 한국에서는 1990년 중반부터 서해안 지역을 중심으로 많이 발견되고 있다.

It is mainly discovered in Southeast and East Asia. Both the male and female of this species have a small horn on their heads while the former has a dent at the center of its thorax. Its black body is subtly glossy with some stripes of stipples on it. In South Korea, it has been discovered since the mid 1990s, mainly in the west coast.

[Distribution] Far East and Southeast Asia (Korea, China, Japan, Taiwan, etc.)

EPISODE
Photograph Book Series of the World Insects Vol. II

Chiang Rai valley, N. Thailand. Dec 16, 2007.

20여 년에 걸쳐 세계를 누비며 채집한
곤충 만큼이나 소중한 추억들을 회상한다.

병을 얻어 몸을 가눌 수 없어
비틀거릴때 마다 누울 곳을 말없이
마련해 주고 음식까지 내어주던
소박한 눈빛과 수줍은 웃음들.

혹여나 허송세월하는 것이 아닌가
자책할 때면 고립된 순수를 머금은 이슬을
내뱉으며 내 눈과 가슴을 일깨워 주던
낯설은 꽃잎들.

나눌수 없는 기억들이지만 적어도
이 한켠에는 몇장의 추억들을 진열해
보고자 한다.

골든 트라이앵글

치앙라이 라오족의 가옥
House of Lao People, Chiang Rai,
N. Thailand. Dec 14. 2007.

치앙라이 라오족의 뱀부보트
Bamboo-boat of Lao People, Chiang Rai,
N. Thailand. Dec 14. 2007.

치앙라이 라오족의 가옥
House of Lao People, Chiang Rai, N. Thailand. Dec 14. 2007.

I look back on memories as precious as insects that I have collected all over the world for about 20 years. People with clean eyes and shy smile who provide a sleeping place and food when I was sick and staggered, Strange petals that drop lonely and pure dew on me to open my eyes and mind when I blamed myself for wasting time, I would like to display some memories on this page although I can not share all the memories.

Golden Triangle, N. Thailand

Photograph Book Series of the World Insects

Photograph Book Series of the World Insects

Photograph Book Series of the World Insects

Photograph Book Series of the World Insects

Photograph Book Series of the World Insects

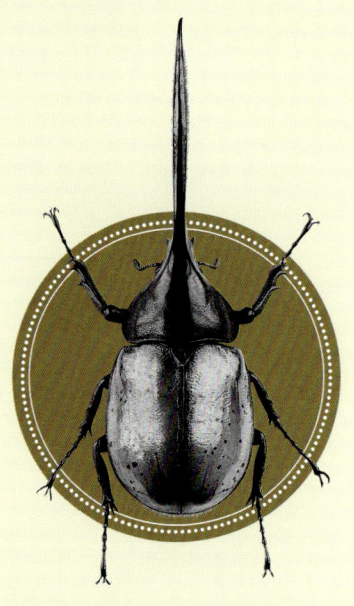

"It is not the strongest of the species that survives,
nor the most intelligent, but the most responsive to change,"
Charles Robert Darwin (1809-1882)

펴낸이
박철영
(주)커뮤니케이션 열림 경기도 파주시 교하읍 문발리 파주출판도시 514-7
TEL. 031)955-0123 FAX. 031)955-0119
www.comopen.co.kr

2009년 4월 20일 인쇄
2009년 4월 28일 발행

Published by
Park, Cheolyoung
Communication Yeollim Co., Ltd.
514-7, Paju Book City, Munbal-ri, Gyoha-eup, Paju-si, Gyeonggi-do,
SEOUL 413-756 KOREA.
TEL. +82-31-955-0123 FAX. +82-31-955-0119

First printed in 2009
First published in 2009

저자(사진, 글)
손민우

Author(Photograph & write)
Son, Minwoo

ISBN 978-89-93849-01-1-76490
ISBN 978-89-959228-9-7-76490(세트)

• 이 책은 저작권법에 의해 보호를 받는 저작물이므로 글과 사진의 무단전제와 복제를 금합니다.
ⓒ All rights reserved. No part of this book may be used or
reproduced in any manner whatsoever without written permission.

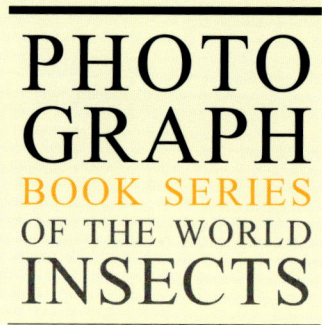